Curious Creatures

ELEPHANT SEALS

GAIL TER

BLACK RABBIT BOOKS

Bolt is published by Black Rabbit Books
P.O. Box 227, Mankato, Minnesota, 56002
www.blackrabbitbooks.com

Marysa Storm, editor; Michael Sellner, designer and photo researcher

Library of Congress Cataloging-in-Publication Data
Names: Terp, Gail, 1951- author.
Title: Elephant seals / by Gail Terp.
Description: Mankato, Minnesota : Black Rabbit Books, [2023] | Series: Bolt. Curious creatures | Includes bibliographical references and index. | Audience: Ages 8-12 | Audience: Grades 4-6 | Summary: "Elephant seals are animals unlike any other, and curious and reluctant readers alike will be captivated by these exceptional creatures' life cycle, habitats, diet, and threats to survival through carefully leveled text, detailed infographics that emphasize visual literacy, and engaging, colorful photography"- Provided by publisher.
Identifiers: LCCN 2020016578 (print) | LCCN 2020016579 (ebook) | ISBN 9781623105655 (hardcover) | ISBN 9781644664902 (paperback) | ISBN 9781623105716 (ebook)
Subjects: LCSH: Elephant seals–Juvenile literature.
Classification: LCC QL737.P64 T475 2023 (print) | LCC QL737.P64 (ebook) | DDC 599.79/4-dc23
LC record available at https://lccn.loc.gov/2020016578
LC ebook record available at https://lccn.loc.gov/2020016579

Image Credits

123RF: nbiebach, 22–23; Alamy: agefotostock, 18; David Osborn, 6; Justin Hofman, 13; Nerthuz, 25; Blue Planet Archive: Carlos Eyles, 12–13; Howard Hall, 23; Jon Cornforth, 6; Dreamstime: Hotshotsworldwide, 1; Palych, 14–15; Swan555, 21; Getty: Ai Angel Gentel, 18; Paul Souders, 11; Royal Geographical Society, 26; iStock: Gerald Corsi, 4–5; flyingrussian, 21; Minden Pictures: Alex Hyde, 3, 31; David Tipling, 25; Gabriel Rojo, 29; Yva Momatiuk and John Eastcott, 8–9, 21, 32; Shutterstock: Brilliance stock, 25; Christian Musat, 25; Gerald Robert Fischer, 10–11; Ivana Kontic, 20; JeremyRichards, Cover; JIAN HONGYAN, 25; Jiang Zhongyan, 25; Ondrej Prosicky, 16–17; Sheila Fitzgerald, 19; Viktorya170377, 14–15, 20

Contents

CHAPTER 1

An Elephant Seal in Action

A male elephant seal lies on a beach. Females relax nearby. The male is in charge. It's the beachmaster. Another male comes near. It wants to take over. The beachmaster rises and roars. Both males lift their chests high, mouths open wide. The beachmaster bites the other male. The challenger fights back but fails. Defeated, it leaves. The master returns to its females.

COMPARING SIZES

LENGTH 10 to more than 13 feet
(3 to 4 meters)

LENGTH 10 to 20 feet
(3 to 6 m)

A Trunklike Nose

Elephant seals are known for their huge noses. They are so big, they look like elephant trunks. These noses are what gives the animals their name. Only males have the trunklike noses.

These mammals are the world's largest seals. Males can be three times larger than females.

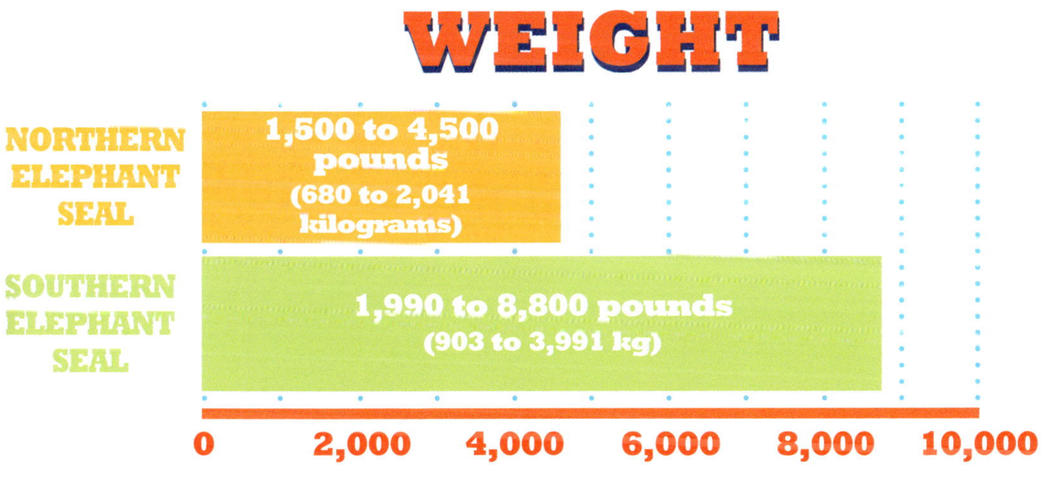

PARTS OF AN ELEPHANT SEAL

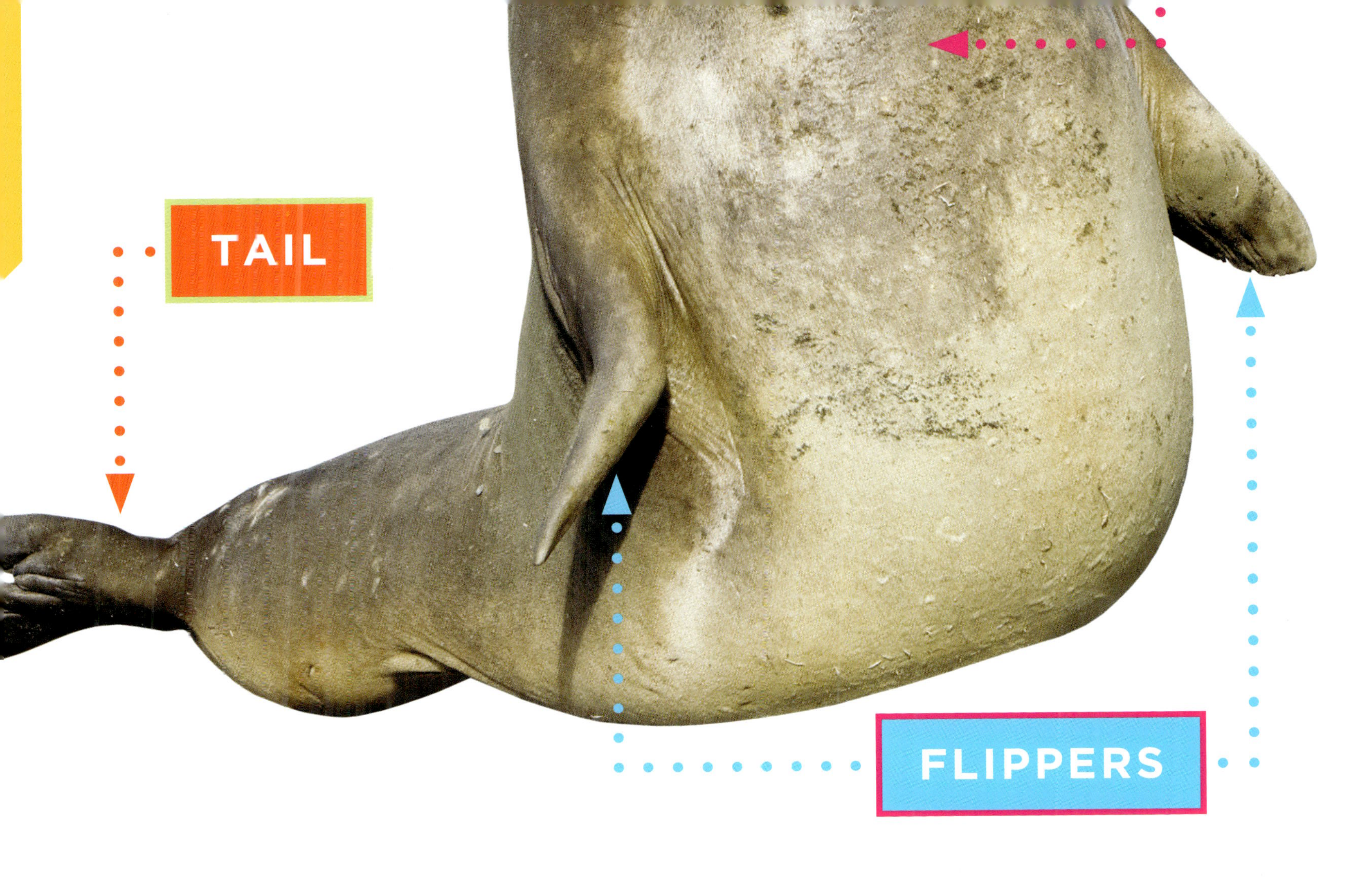
TAIL
FLIPPERS

CHAPTER 2

FOOD and Home

Elephant seals find all their food at sea. Most often, they eat squid. They also hunt fish, octopuses, and small sharks. These seals can dive deep to find **prey**.

Not all elephant seals hunt in the same places. Females tend to hunt in the open sea. Males stay closer to land. Scientists think they hunt in different places because of their sizes. Killer whales also hunt close to land. Large males stand a better chance against these **predators**.

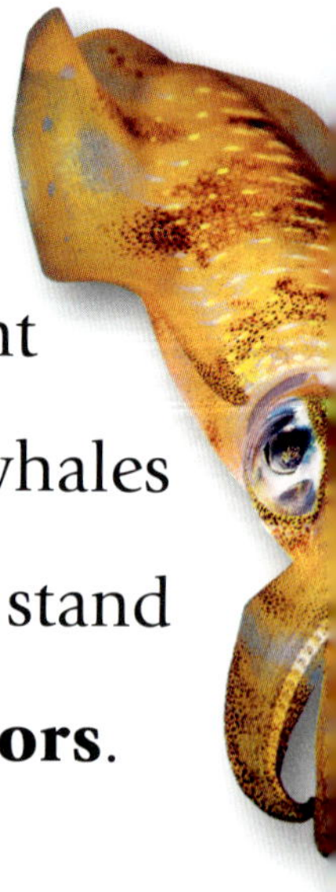

COMPARING MAXIMUM RECORDED DIVING DEPTHS

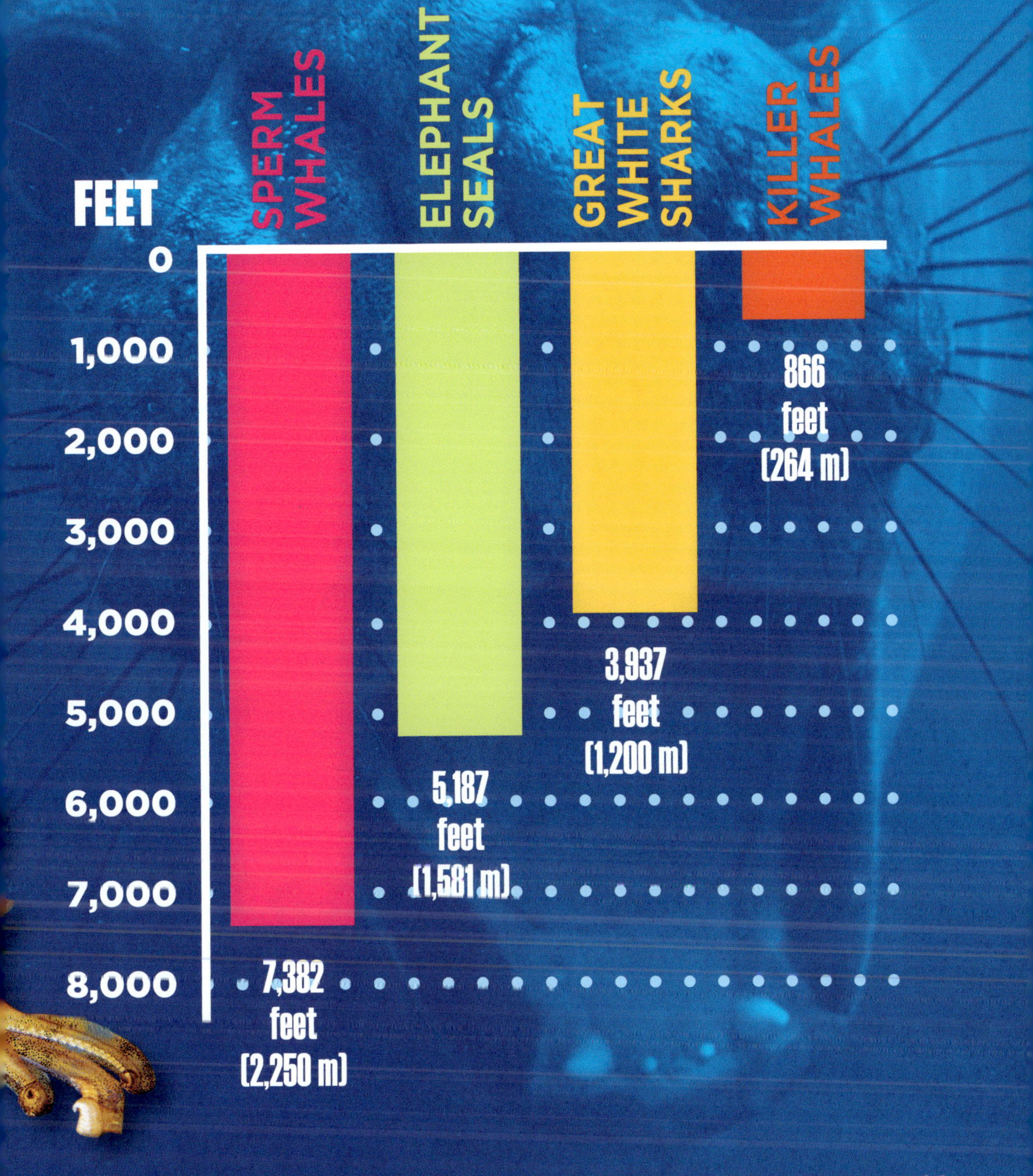

Life at Sea

Northern elephant seals live off the west coast of North America. Southern elephant seals live near Antarctica. Both types spend eight to 10 months of the year at sea. The water they swim in can be very cold. Thick layers of **blubber** keep them toasty warm.

WHERE ELEPHANT SEALS LIVE

Elephant Seal Range Map

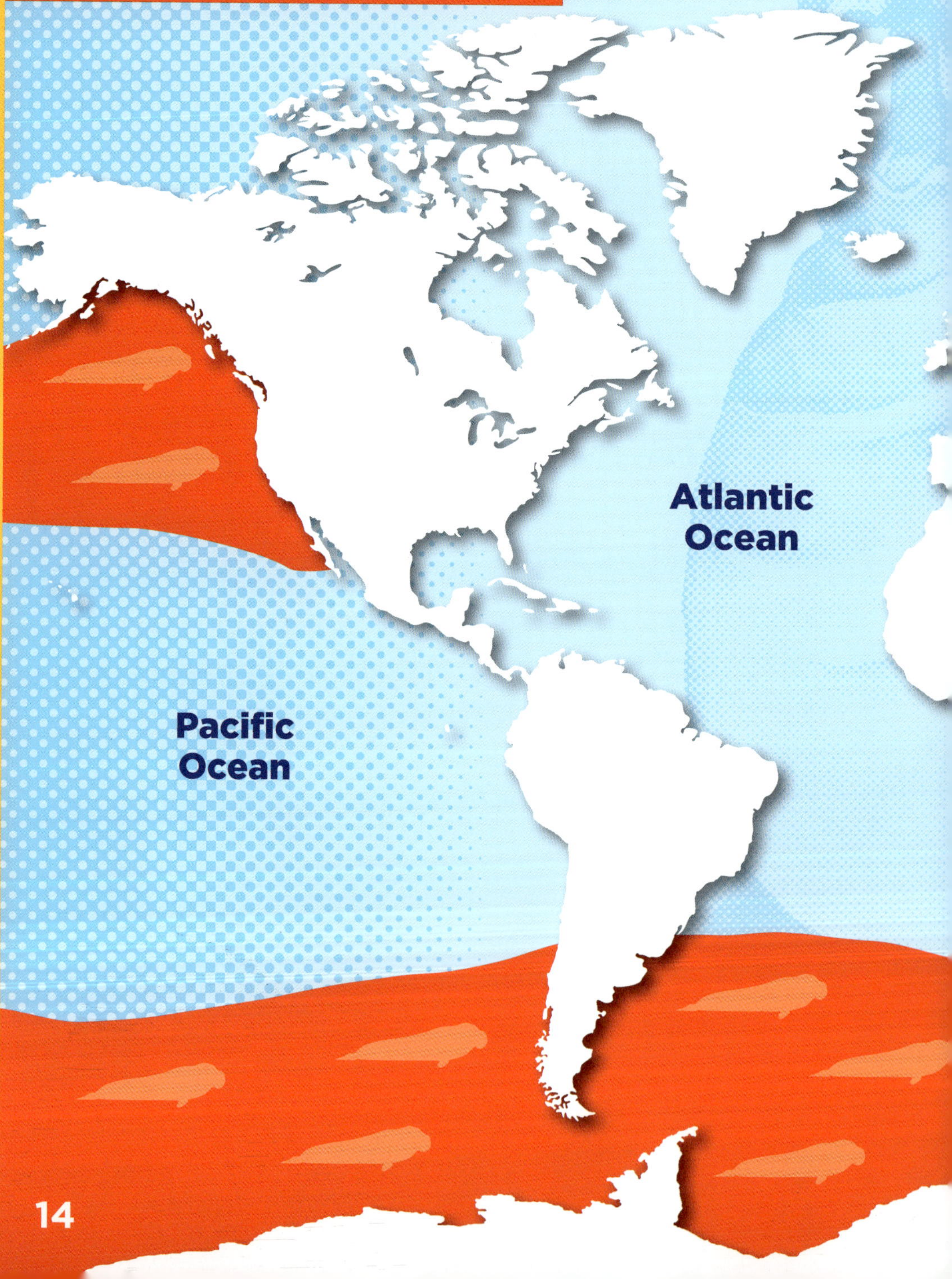

Arctic Ocean
Pacific Ocean
Indian Ocean
Southern Ocean

CHAPTER 3

The Family

When at sea, elephant seals are mostly alone. But twice a year, they meet on land. The first time is to give birth and to **mate**. The second time is to **molt**. When molting, they lose sheets of fur and skin.

While on land, adult elephant seals do not eat. They live off their blubber.

Becoming a Top Male

FIRST ON LAND

FIGHTING FOR CONTROL

WATCHING OVER FEMALES

Time to Mate

Males are ready to mate by about age six. During mating season, they are the first to go on land. They fight to figure out which males are **dominant**. The top males get to be beachmasters and have groups of females. They mate with all the females in the group. Other males do not get to mate.

Pups are Born

When a female comes on land, it does not mate right away. First, it gives birth to one pup. Then it mates. It also nurses its pup. After four weeks, the mother goes back to the sea. The pup must take care of itself. The pup stays on land for about eight more weeks. It grows and learns to swim. Then it's time to begin its life at sea.

COMPARING WEIGHTS

NORTHERN ELEPHANT SEAL NEWBORN	NORTHERN ELEPHANT SEAL ONE MONTH OLD	NORTHERN ELEPHANT SEAL ADULT FEMALE
ABOUT 75 POUNDS (34 kg)	324 POUNDS (147 kg)	2,000 POUNDS (907 kg)

pup

nursing

swimming

By the Numbers

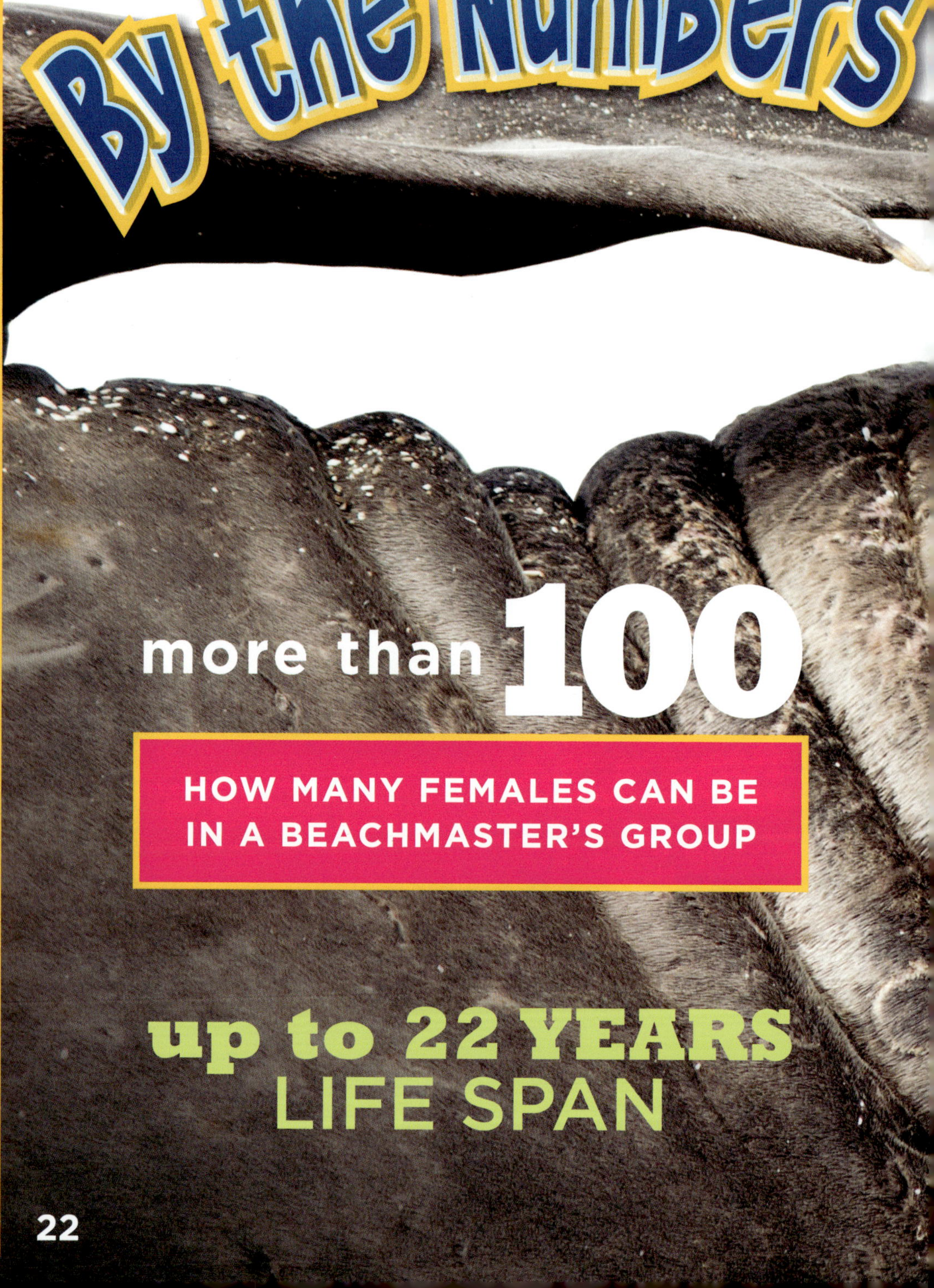

more than **100**

HOW MANY FEMALES CAN BE IN A BEACHMASTER'S GROUP

up to 22 YEARS
LIFE SPAN

10 **POUNDS** **(4.5 KG)**

HOW MUCH WEIGHT A PUP CAN GAIN IN A DAY

30

TEETH

UP TO 2 HOURS how long they can stay under water

CHAPTER 4

Danger!

As adults, elephant seals have few predators. But great white sharks and killer whales sometimes eat them. The seals escape most attacks because they are so big. They can dive deeper than their predators too.

Elephant Seal Food Chain

This food chain shows what eats elephant seals. It also shows what elephant seals eat.

Elephant Seals and Humans

Humans have caused big problems for these giant seals. In the 1800s, people killed most of the seals. They hunted the seals mainly for their blubber. They turned the blubber into **oil**. The seals almost became **extinct**. Now, laws protect the seals. They are no longer endangered.

Helping Research

Climate change affects people and animals around the world. Scientists have turned to the seals for help. They tag the seals with data trackers. The seals take these sensors where scientists can't go and gather water data. They measure the water's temperature. They also measure depth. Over time, scientists will use this data to fight climate change. Thank you, elephant seals!

Elephant seals spend about 90 percent of their lives in water. They're perfect for ocean research.

blubber (BLUB-uhr)—the fat on whales and other large sea mammals

climate change (KLAHY-mit CHEYNJ)—changes in the earth's weather patterns

dominant (DOM-uh-nuhnt)—controlling or being more powerful than all others

extinct (ek-STINGKT)—no longer existing

mate (MAYT)—to join together to produce young

molt (MOLT)—to lose a covering of hair, feathers, or skin and replace it with new growth

oil (OYL)—a thick, greasy liquid that burns easily

predator (PRED-uh-tuhr)—an animal that eats other animals

prey (PRAY)—an animal hunted or killed for food

BOOKS

Chang, Kirsten. *Seal or Sea Lion?* Spotting Differences. Minneapolis: Bellwether Media, 2021.

Levy, Janey. *The Beastly Elephant Seal.* Nature's Freak Show: Ugly Beasts. New York: Gareth Stevens Publishing, 2020.

Murray, Julie. *Animals in Antarctica.* Animal Habitats. Minneapolis: Abdo Kids Junior, an imprint of Abdo Kids, 2021.

WEBSITES

Elephant Seals
www.nationalgeographic.com/animals/mammals/group/elephant-seals/

Elephant Seals
www.nps.gov/pore/learn/nature/elephant_seals.htm

Southern Elephant Seals
oceanwide-expeditions.com/to-do/wildlife/elephant-seal

INDEX